CATS

CATS

In from
the Wild

By
Caroline
Arnold

photographs
by
Richard
R.
Hewett

Carolrhoda Books, Inc.
Minneapolis

We are grateful to many people for making it possible for us to take the photographs for this book. In particular, we thank Sarah Paul; Ronald Park; the Los Angeles Zoo; the Los Angeles County Natural History Museum; Marine World Africa USA, in Vallejo, California; Wildlife Safari, in Winston, Oregon; and the Rosicrucian Egyptian Museum, in San Jose, California. We also give special thanks to Walter Lynd and Carolyn Perry for their generosity in letting us get to know and photograph their cats and kittens. We also want to recognize our own cats, Tama Neiko, Muffin, and Abigail, who submitted to being photographed for this book and who have been loving and faithful companions for many years.

Photograph on page 20 courtesy of Matthew Arnold
Words that appear in **bold** type are listed in the glossary on page 46.

Ruth Berman, Series Editor
Zachary Marell, Series Designer

Library of Congress Cataloging-in-Publication Data

Arnold, Caroline.
    Cats : in from the wild / by Caroline Arnold ; photographs
by Richard R. Hewett.
        p.   cm.
    Includes index.
    Summary: Discusses the physical characteristics and behavior of
cats since their evolution from ancient cats forty million years ago.
    ISBN 0-87614-692-2
    1. Cats—Juvenile literature. 2. Felidae—Juvenile literature.
3. Cats—Behavior—Juvenile literature. 4. Felidae—Behavior—
Juvenile literature. [1. Cats.] I. Hewett, Richard, ill. II. Title.
SF445.7.A76 1993
636.8—dc20   92-32986
                                                                CIP
                                                                AC

Manufactured in the United States of America

1   2   3   4   5   6   98   97   96   95   94   93

636.8
ARN
c.1

9 - 93

$19.95

Contents

A cat depends on stealth to capture its prey.

Chapter 1

What is a Cat?

E yes bright, whiskers quivering, and supple body ready to spring, a domestic cat stalks a bird on a city street. Far away in the jungles of India, a hungry Bengal tiger uses a similar technique to catch its meal. After sighting their prey, both animals crouch low and slowly begin their approach, trying to stay hidden for as long as possible. Then, with eyes and ears alert, tails twitching in anticipation, they creep forward on padded paws. Finally, at the last moment, they spring to attack.

THE CAT FAMILY

Most of us will never see tigers in the wild. But if you watch your house cat carefully, you can see how its body and much of its behavior are like that of its wild relatives. Both domestic cats and tigers are members of the cat family, and although

7

Left: *A saber-toothed cub had to be careful when playing with its mother.*

Right: *A North American saber-toothed cat called Smilodon once hunted ice age mammals in Southern California. It became extinct about 10,000 years ago.*

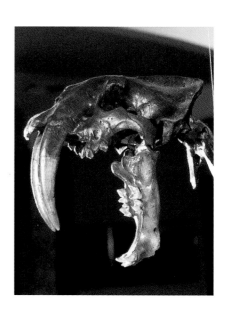

Sharp claws help cats to be excellent climbers, although going up is sometimes easier than coming down.

each has adapted to its own way of life they share many characteristics. Like other cats they are furry, have sharp teeth and claws that can be **sheathed**, or pulled in. Both have supple, muscular bodies with quick reflexes that help them to be excellent hunters. Cats are among the most beautiful and graceful creatures alive and are loved by people all over the world.

The first cats appeared on earth about 40 million years ago and were descendants of early carnivores, or meat-eaters, called miacids. These ancient cats developed into two sub-families. One of these subfamilies is the now-extinct group called the Machairodontinae, or saber-toothed cats. Each saber-toothed cat had a pair of enormously elongated teeth in the front of its mouth. Scientists think that the saber-toothed cat may have used these teeth to stab its prey or to rip open an animal's soft underbelly.

The second subfamily of ancient cats is the Felidae, the family that includes all modern cats. Scientists divide the Felidae family into three groups according to shared characteristics. Each of these large groups is called a **genus**. The first genus, *Panthera*, includes six species of large cats: the tiger, lion, leopard, snow leopard, clouded leopard, and jaguar. The *Acinonyx* genus has only one member, the cheetah. The remaining cats, which include domestic cats, bobcats, ocelots, and many others, are part of the *Felis* genus. Most members of this genus are small or medium-sized, except for the cougar, which can grow to be up to 200 pounds.

There are approximately 35 species of cats, ranging in size from the tiger, which can weigh more than 800 pounds, to the tiny black-footed cat, which weighs about 5.5 pounds. The total number of cat species varies slightly because some experts group several subspecies together as one species, and others classify them as separate species. Usually, members of different species cannot mate and produce offspring. However, closely related animals do sometimes reproduce successfully. For instance, in zoos, lions and tigers have been known to crossbreed. The offspring are known as ligers or tions.

Although tigers are usually orange with black stripes, white tigers are sometimes born.

The black-footed cat is a close relative of the African wildcat and is noted for its fierceness.

Compared to other domestic animals, the cat is the most like its wild relatives, both in its physical appearance and in its behavior. Experts think that among the present-day domestic cats, the Abyssinian breed probably looks most like the cats that lived in ancient Egypt.

By keeping vermin such as rats and mice under control in ancient Egypt, cats became valuable to people.

THE FIRST DOMESTIC CATS

The ancestors of most domestic animals, such as dogs, cows, or horses, were wild animals that had been caught by people and **tamed** for human use. After many generations of being bred in captivity, the animals became **domesticated**, having adapted to living and working with people. Cats, on the other hand, joined people of their own accord. They chose to live near people but continued to hunt and interact with each other much as they had done in the wild. Even as pets, cats tend to retain some of their independent ways.

Most people believe that the domestic cat is descended from an animal called the African wildcat, a small striped cat similar in appearance to the domestic tabby. About 6,000 years ago in ancient Egypt, some of these small native cats began to live near farms and villages. The wildcats had discovered that there were plenty of rats and mice to eat near the places people stored their grain. The Egyptians welcomed the cats because they helped to get rid of the pesty rodents that were destroying their food. The cats gradually adapted to life with people and were considered domesticated.

Left: *Wooden statue of a cat from ancient Egypt*

Right: *Mummies of Egyptian cats were buried in special cat cemeteries.*

Cats were so valuable to the ancient Egyptians that they were worshipped as gods and goddesses. One goddess had the head of a cat and the body of a woman. Her name, Bast or Pasht, may be the origin of our word "puss." In the city of Bubastis, the Egyptians built the Great Cat Temple in which archaeologists have found thousands of cat mummies.

The use of cats as pets and to control pests slowly spread from ancient Egypt to the Middle East, India, China, Japan, and Europe. Although cats continued to be valued for their ability to hunt mice, they never again achieved the high status they had held in Egypt. In some countries, such as China, domestic cats became sources of food. During the Middle Ages in Europe, the domestic cat was associated with evil and often persecuted. Some people still believe that a black cat brings bad luck. On the other hand, some people believe a black cat is the symbol of good luck.

This pre-Columbian jaguar statue from Mexico is decorated with spots similar to those of a living jaguar.

Like all modern cats, the domestic cat is a member of the Felidae family.

All cats have powerful muscles in their hind quarters that they use for running and jumping. A domestic cat can run at speeds up to 30 mph over short distances and can easily jump up to five times its own height.

Today, domestic cats can be found almost everywhere and are not much different from their ancestors, the wildcats. Several subspecies of wildcats are found in parts of Africa, Europe, and Asia, and in some cases have been known to mate with domestic cats.

The small domestic cat was not the only species to be used by people. Some larger species were also tamed. Cheetahs were often trained as hunting animals by royalty in the Middle East and India, and jaguars were kept as pets by the ancient Aztecs in Mexico.

THE CAT'S BODY

All cats are carnivores, or meat-eaters, and every part of a cat's body helps it to be an efficient hunter. Compared to other animals, a cat has a relatively small head in relation to the rest of its body. This helps it to fit through tight places. A narrow, flexible body and a good sense of balance allow a cat to walk easily through dense shrubbery or along a narrow ledge in pursuit of prey.

Although cats are remarkable climbers, they do sometimes fall. However, even when cats fall from relatively high places, they often escape unharmed because cats have the ability to turn themselves around in midair, so that they land on all four feet. Their elastic muscles and skeletons help their bodies to absorb the impact without getting hurt.

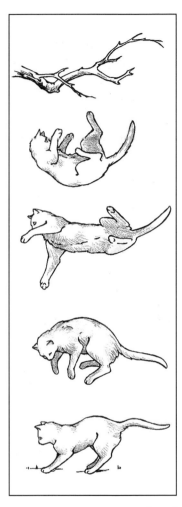

Above: *A cat turns its head first, then its body, when righting itself in midair.*

Left: *A cat's long, flexible tail is important in helping the cat keep its balance. The tail may have as many as 28 small bones called vertebrae.*

Left: *Cheetahs are the only members of the cat family who cannot retract their claws.*

Right: *A cat has needle-sharp claws.*

Above: *Muscles inside the cat's foot allow the claws to be extended when needed.*

Most cats are stealthy hunters, approaching their prey slowly and silently on padded paws. A cat has four toes on its back feet and five toes in front, each ending in sharp claws. Cats (except cheetahs) are among the few creatures that have claws that can be sheathed when not in use. This helps them to walk silently on hard surfaces and to keep their claws sharp. The claws can be extended for defense, to catch and secure prey, or to grab hold when climbing.

As a cat stalks its prey, it waits until it is close enough to spring. Then it leaps forward, trying to surprise and capture the animal. To kill its prey, a cat grabs it with sharp claws and then bites it on the neck. Small cats eat rodents and other small mammals, birds, reptiles, and sometimes fish. Because of their great size and strength, the large cats are able to kill big animals such as antelope, zebra, or wild goats.

A cat's teeth, like those of other carnivores, are uniquely adapted for killing and eating prey. An adult cat has 30 teeth —12 incisors, or cutting teeth, in the front of the mouth; 4 canines behind these; and 14 molars in the back of the mouth. Plant-eating animals, such as sheep and cattle, use their upper premolars and first lower molars for chewing. In the cat, however, these are sharp cutting teeth, called carnassials, which work against each other like scissors to cut and slice food. All carnivores have carnassials, but in cats they are particularly sharp.

By curling the end and sides of its tongue, a cat scoops liquids into its mouth.

Cheetah. When a cat eats, it cuts off pieces of meat with the sharp teeth at the side of its mouth.

A cat uses its rough tongue to clean itself and to remove meat from bones.

Cats cannot easily move their jaws from side to side, which means they cannot chew their food well before swallowing it. This is not a problem because their strong stomach juices can digest large pieces of food and even break down pieces of bone that might have been swallowed.

THE CAT'S SENSES

All of a cat's senses are keenly developed for hunting. A cat has excellent vision, and at night it can see at least six times better than a human. Although a cat cannot see colors well, it can see shapes and movement as long as there is a little light.

Many cats hunt at night since this is when prey animals such as mice and rats are active. Sometimes when you see a cat at night, its eyes seem to shine in the dark. This is because the cat's eyes contain a substance called **tapetum lucidum** (Latin words meaning "bright carpet"). The tapetum lucidum reflects light to the back of the eye and helps the cat to see better in dim light.

When light is low, a cat's pupils open wide.

In bright light, the pupils become narrow slits.

Excellent vision and keen hearing help the bobcat hunt the birds, rabbits, and rodents that are its food.

The caracal's large ears help it to hear especially well. All cats can move their ears independently of each other, which is one reason their sense of hearing is so acute.

A cat also has a good sense of hearing and can turn its ears to catch a variety of sounds. A cat can hear a much wider range of sounds than we can. Its ears are particularly sensitive to **ultrasounds**, that is, sounds above the range of human hearing. (A "silent" dog whistle produces ultrasounds that dogs and cats can hear but people can't.) Tiny, high-pitched squeaks in the ultrasound range are often made by the rodents that cats eat, and a cat can use these sounds to locate its prey. House cats are often attracted to the noises of crumpling paper or jangling keys because they make similar high-pitched sounds. A cat's sense of hearing also helps it to detect approaching danger.

A cat's sense of smell is not as highly developed as its senses of vision and hearing. Nevertheless, odors are important to cats. A cat detects odors both through its nose and with the **Jacobson's organ** located on the roof of its mouth. Sometimes you may see a cat pulling its lips up and backward in a kind of grimace. This movement, which is known by the German word *flehmen*, helps to pull air into the mouth and past the Jacobson's organ. A cat identifies the presence of other animals by each animal's special odor.

A cat's sense of touch is aided by sensitive whiskers on either side of its nose. The scientific name for these long, stiff hairs is **vibrissae**. Although the whiskers beside its nose are the most important, the cat also has a few vibrissae above its eyes, on its cheeks, and behind its front legs. The cat uses its whiskers to feel the space between objects, which helps it to know whether or not its body will fit through a gap. The whiskers can also feel very slight air movements. Even in the dark a cat can feel the air move around objects, so it can avoid bumping into them.

The flehmen *response enhances the cat's ability to detect odors.*

A domestic cat usually has 12 whiskers on either side of its nose, each one deeply embedded into the skin. They convey important information to the cat about its surroundings.

A domestic cat surveys its territory.

Chapter 2

Cat Behavior

C ats are noted for their graceful movements and stealthy habits, but these are just some of a wide variety of cat behaviors. As with all animals, behavior can help or hinder a cat's chance for survival. Patrolling and marking territory, fighting, mating, giving birth, and rearing kittens are all part of a cat's behavior. Much of the cat's day is also spent finding food, eating, sleeping, and keeping clean.

A CAT'S TERRITORY

In general, cats are solitary creatures. Each cat usually lives in its own territory, which it defends against other cats. Territories of males are usually much larger than those of females. For domestic cats in the country, territories may be up to 150 acres for males and 15 acres for females. In the city, however, cat territories are usually much smaller because there are so many more cats. However, even a well-fed house cat retains its territorial **instincts,** and stakes out its turf in its backyard or even within a house.

One of the ways that cats mark the boundaries of their territories is by scratching. Except for cheetahs, who cannot retract their claws, cats frequently sharpen their claws by scratching, which removes snags and loose scales. At the same time, the marks a cat leaves behind are a visual signal to other cats that it has been there. Also, when a cat scratches, **scent glands** in the pads of its paws mark the spot, leaving behind a personal odor that other cats can detect.

Another way cats mark the boundaries of their territories is by spraying them with urine. To spray, a cat backs up to its target, which could be a bush, wall, or post, and aims a stream of urine backward onto it. Although both male and female cats

A cat cautiously looks over the situation before going out.

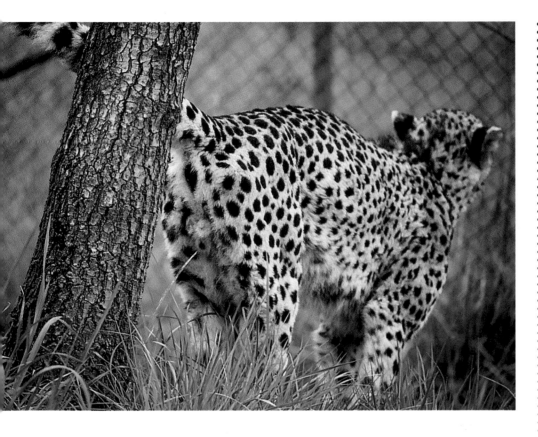

A cheetah backs up and raises its tail as it prepares to spray urine on a tree to mark its territory.

spray, females do it less often, and their urine does not have the same strong odor that a male's has. A male cat has special scent glands near his tail that produce a strong smell in his urine. This smell is quite unpleasant to humans but conveys important information to other cats. Every cat has a unique odor, and any cat that passes by a marked area can smell it and know who has been there before. The strength of the odor also indicates how long it has been since it was sprayed.

A cat makes frequent tours of its territory to see what is going on and to check the various scent posts. Even though a domestic cat's territory may be small, it likes to check its territory regularly. This is why cats frequently beg to be let out and then ask to be let in again a few minutes later.

A friendly cat rubs a person's leg in greeting.

Adult cats usually keep a respectful distance from each other, even when they live together.

When territories are large, cats can usually avoid meeting each other. When a cat does encounter another creature, its reaction depends on who the person or animal may be. In one form of the friendly greeting, a cat rolls over on its back and twitches its tail. In this position the cat is quite defenseless, so it only does this in the presence of people or animals that it really trusts. Another friendly greeting involves repeated rubbing, especially along the side of the mouth and the top of the head. A cat has scent glands on its cheeks, head, and feet. When a cat twines itself around your legs, it is leaving some of its scent on you, but it is also picking up some of your scent, which helps the cat to confirm your identity.

Several cats living in the same household usually learn to tolerate each other and sometimes become close companions.

Two jaguars snarl at each other. Jaguars are spotted cats similar to leopards but with larger, heavier bodies. Most have reddish coats with dark spots; those that appear all black actually have black spots against a dark background.

However, in a neighborhood with a number of cats, repeated encounters determine a **dominance hierarchy**, and those cats with more aggressive personalities establish themselves as the **dominant** animals. When a cat meets a rival, its reaction depends on whether it is the stronger or weaker individual in that situation. A dominant cat will threaten to attack by advancing sideways toward the other animal with its back raised; meanwhile, the **subordinate** animal crouches in a defensive position. Although there may be growling and snarling, there is usually no actual fight because the weaker animal eventually turns and runs, conceding the challenge to its opponent. Thus, both cats avoid what could be a life-threatening fight. When a cat fight ends in a physical attack, each cat claws at the other and tries to bite the other's neck.

This puma has assumed the defensive position of a subordinate cat.

A cat defends itself with sharp teeth and claws.

When a cat meets a large, threatening animal, such as a dog, it may either run for safety or try to defend itself. A defensive cat hisses and arches its back to make itself look bigger. If the other animal comes too close, the cat will swipe its nose with sharp claws. Most dogs learn quickly to avoid hissing cats.

Despite their usual preference for being alone, domestic cats do sometimes live in groups, especially if they have grown up together and if there is plenty of food available. Studies of farm cats show that groups of females often build their nests together and may even share nursing duties in much the same way that female lions in a **pride** look after each other's cubs. For social interaction within the group, kittens and cubs take on the dominance rank of their mothers.

MATING AND REPRODUCTION

When a female cat is ready and able to mate, she is said to be in heat. A domestic cat usually has two or three heat periods a year, each lasting about 10 days. You can tell that a female cat is in heat because she tends to eat less, be restless, meow

The cat's posture warns the dog to keep its distance.

more, and rub her head against objects and people she knows. When the heat peaks, usually on the third day, the cat yowls, writhes on the floor, and frequently licks her paws and rear end. If you touch her at this time, she will raise her rear end and crouch down in front, assuming the position for mating.

Male lions are distinguished by handsome manes. Lions are the only cat species in which males and females are markedly different from each other.

A male cat can mate at any time of the year. His territory usually overlaps with those of several females, and as he patrols this space, he checks to see if any female cats are ready to mate. He can detect when a female in his neighborhood is in heat by her odor and her yowling. Often several males will gather around a female in heat and fight with each other for the privilege of mating with her. The male cats also yowl and may spray urine. When the female's heat is over, the males lose interest in her and go their separate ways. In most cat species, only the mother cat cares for the offspring.

After a successful mating, a female domestic cat is pregnant for about nine weeks. (The length of pregnancy varies from 57 to 70 days.) During this time, one to eight tiny kittens are developing inside her uterus, each one enveloped in its own birth sac. By the time the kittens are ready to be born, the female's belly bulges noticeably.

As other cat species do when they mate, the male jaguar grasps the female with his front paws and bites her at the back of the neck.

About two weeks before she is due to give birth, a pregnant cat begins to look for a good place for her nest. An outdoor cat will find a secluded place in a shed, hayloft, or under thick bushes. A cat that lives with people will usually accept a sturdy box in a quiet corner.

Immediately after each kitten is born, the mother cat licks it to remove the birth sac. Then she bites off the umbilical cord and eats the afterbirth. This keeps the nest clean, provides nourishment to the mother, and—for cats in the wild—it removes material that might smell and attract predators. The entire birth process may last several hours or longer depending on the number of kittens and how quickly each

A mother cat usually prefers a nest where her kittens are hidden, so predators cannot find them easily.

Kittens begin to purr when they are about a week old.

When she is in heat, a female may mate several times. If she mates with more than one male, her kittens will have different fathers. That is one reason kittens in the same litter sometimes look quite different from one another.

one is born. At birth, each domestic kitten is about 5 inches long and weighs between 2 and 4 ounces.

As soon as a newborn kitten is licked clean, it begins its search for one of the teats on its mother's belly. It cannot see or hear at birth, but it can find its way by using its senses of touch and smell. When it finds a teat, it takes hold with its mouth and begins to suck milk.

The mother cat normally has eight teats lined up in two rows of four each. Each kitten soon finds a favorite teat and uses it almost exclusively during the weeks that it nurses. If any of its littermates tries to suck there, the kitten pushes it away. When the nursing kittens suck, they knead their paws against their mother's belly. The kneading helps to make her milk flow. A teat provides milk as long as it is used regularly. When a kitten is weaned, the milk gland dries up.

Kittens are able to purr and suck at the same time. These gentle noises assure the mother cat that her kittens are

content and all is well. At the same time, the mother cat may respond with her own purring, which lets the kittens know she is relaxed too. Adult cats also purr as a greeting and to let others know that they are in a friendly mood.

How cats purr has always been somewhat mysterious. Many scientists believe that purring results from the vibration of blood in a large vein in the cat's chest. All the small cats purr, and they can do so even when breathing or eating. Of the big cats, only cougars, clouded leopards, and snow leopards are able to purr.

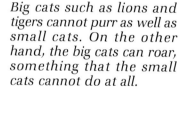

Big cats such as lions and tigers cannot purr as well as small cats. On the other hand, the big cats can roar, something that the small cats cannot do at all.

A mother cat nurses her four tiny kittens.

Two-week-old kitten

Four-week-old kitten

Six-week-old kittens

If a kitten wanders too far and gets lost, a plaintive meow lets its mother know where to find it.

Young kittens grow rapidly. Their eyes open between their 8th and 11th day. At two weeks, they begin to crawl around the nest. By the time they are four weeks old, they start to play with each other and to chase their mother's tail. At first kittens are clumsy, but they gradually become more adept. Although it appears that kittens' play is just for fun, it is important in helping them to develop their coordination and to sharpen their reflexes. As kittens become more independent, they begin to explore away from the nest.

Like wild cats, domestic cats who do not have human caretakers must learn how to find their own food. They begin this process at an early age. During the period that kittens are nursing, their mother leaves them periodically to find her own food. At first she brings food back to the nest and lets the kittens watch her eat it.

Kittens grow baby teeth during their first eight weeks of life; these are then replaced with a full set of permanent teeth by the time the kitten is seven to nine months old. As the kittens grow older and their teeth become stronger, the mother brings

Like a kitten stalking a string, a tame tiger seems to enjoy a game of tug-of-war with its trainer.

back dead prey and gives it to the kittens to eat. Later on, the mother returns with the prey still alive and allows the kittens to kill it themselves. She waits nearby and if they let it escape, she catches it so they can try again. In this way, the kittens gradually develop their skills as hunters, so that when their mother leaves them to mate and raise a new litter, they will be able to survive on their own.

In the same way that a mother cat brings food to her growing kittens, a house cat, especially if it is a female, may bring a freshly killed mouse or bird to its human caretakers, behaving as if the people were her kittens and needed to be fed.

A mother house cat does not need to teach her kittens how to hunt. Instead, her human caretaker will gradually introduce the kittens to solid foods. Although house cats and wild cats in zoos will never have to hunt like their counterparts in the wild, they exhibit the same behaviors when stalking a string or playing with a toy mouse. Their predatory behaviors are instinctive, that is, they are born with the ability to hunt.

Cats can fall asleep almost anywhere.

A sleepy lion naps in the sun.

Adult cats retain some of their kitten behavior, especially when they are raised as pets. They seem to view their human owners as large mother cats and respond to them in the same way they did to their own mothers when they were dependent on them. A mother cat licks her kittens frequently both to keep them clean and to show her concern for them. Similarly, a grown cat enjoys being stroked by a favorite person. A grown cat may also knead its paws against a person when it is feeling relaxed and content, just as it did while nursing as a kitten.

RESTING

A domestic cat normally sleeps about 16 hours a day. Sleep varies from short catnaps of just a few minutes to longer rests of an hour or more. Cats are champion sleepers and can fall asleep at almost any time. They can afford to spend most of their lives asleep because they are traditionally such efficient hunters. Some carnivores, such as wolves, must spend a lot of time pursuing their prey, but cats usually just wait quietly for their prey to come to them. After they have killed and eaten their meal, cats rest until they are hungry again. Their inactivity helps them to conserve their energy until it is needed.

To clean its head, a cat first licks its paws and then uses them to wash its face and ears.

The long "spines" on a tiger's tongue give it the texture of very coarse sandpaper.

GROOMING

When cats are not sleeping or eating, they spend much of their time washing themselves. A cat's long, rough tongue is covered with tiny, spinelike projections, called **papillae**. This rough surface, which helps a cat lift food to its mouth and lap up liquids, also works like a brush or comb when a cat grooms its fur. Using its teeth to remove burrs or other objects and its tongue to lick the fur smooth and clean, a cat methodically works from front to back while giving itself a bath.

A cat's fur coat helps to keep it warm and protect it from the weather. Although we usually think of felines as being creatures with soft fur, the texture can vary from the coarse hairs of the cheetah's short coat to the silky softness of a purebred angora cat.

The Peke-Face cat has a squashed face like a Pekinese dog.

Chapter 3

Cats and People

Although cats have had a long association with people, it has only been in recent years that we have begun to understand some of their behavior. The more we know about our pets, the better we can relate to them and respond to their needs. Living with a cat is fun, and it also provides the opportunity to observe animal behavior up close. And if you watch carefully, you may be able to catch glimpses of ways your domestic cat is still like its wild relatives.

BREEDS OF CATS

Domestic cats come in many sizes, shapes, and colors. The largest grow to be about 18 pounds (although some grow to be even heavier) and the smallest weigh about 3 pounds; their

The white Persian cat has long, silky fur and orange, blue, or odd-colored eyes.

The delicate Singapura cat was originally imported from Singapore.

bone structure ranges from delicate to stocky; and their fur can be long or short with colors ranging from black to white, including various multi-colored patterns. Yet, despite the wide variety of forms, all domestic cats are one species and can mate with each other.

As with all living things, a cat's physical appearance is determined by tiny structures in its cells called genes. Each gene or group of genes produces a characteristic such as eye color, fur length, or the shape of the bones. The genes are organized in pairs, with half of each pair contributed by each parent. The combination of genes determines the ways an individual is like its mother, its father, or both.

Occasionally a gene changes, or mutates, to produce a new characteristic unlike either that of the mother or the father. The new mutated gene will then appear in the next generation. Some breeds of domestic cat, such as the Scottish fold whose ears bend forward, are the result of genetic mutations. Others, such as the Siamese or Abyssinian, reflect the varied ancestry of wild cats in different parts of the world.

When people purposely breed two animals together to produce a particular characteristic in the offspring, it is called **selective breeding**. If a large number of cats are produced that have the same traits, then an international jury of cat breeders and judges may declare it to be an official breed. They decide how that breed should look and establish a "standard" that is used to judge these animals in a show. Twenty-five breeds are recognized by the Cat Fanciers Association, the largest such group in the United States.

Many people breed cats either as a business or for fun. Usually they belong to cat clubs and exhibit their cats at shows. Each cat is judged on how well it conforms to the standards set for its breed.

There are more than 58 million cats in the United States today. Although some of these are recognized breeds, most are of mixed ancestry. Whether your cat is a purebred or an "alley" cat, it is the same species as all other domestic cats and its behavior will be much the same as theirs.

The Rex cat has a pixie face and large, pointed ears.

At cat shows, judges examine each animal and grade it according to the standards for its breed.

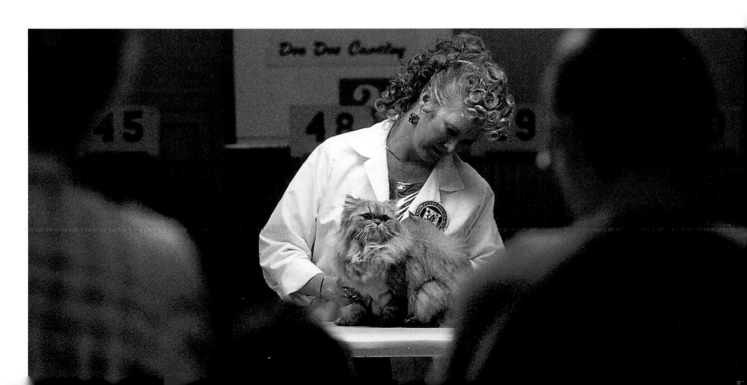

Handling kittens as they grow up helps to get them used to being around people.

CATS AS COMPANIONS

Cats can be excellent companions. They are beautiful to watch and are loving and affectionate. Everyone benefits from the mutual trust that develops between people and their pets. Studies have shown that the act of petting a cat has a soothing effect on both the cat and the person and can help both to relax.

When you keep a cat as a pet, you are responsible for taking care of its daily needs. Every cat needs food and water, places to sleep and exercise, and, if it is kept indoors, a litter pan. For traveling or trips to the vet, a carrying case helps to keep your cat safe and comfortable. Many people like to buy toys for their cats, but an empty spool or a strand of yarn will make a cat just as happy.

Sometimes domestic cats become lost, are abandoned by people, or are otherwise left to fend for themselves. Although they behave like wild cats and no longer depend on humans, they are not considered to be truly wild animals. Any domestic animal that goes back to living in the wild is said to be **feral**. Feral cats can be found all over the world, especially in cities where food is available in dumps and garbage cans.

Although feral cats can usually manage on their own, it is often difficult for them to find enough food. Without human care, they are more likely to become sick or victims of accidents. Too many cats and kittens end up in animal shelters because they do not have homes. Many shelters require that animals be spayed or neutered when being adopted. This helps to reduce the number of unwanted domestic animals. If you have a cat, you should make sure that you care well for it. And if your cat does have kittens, be sure to place them in good homes.

Cats that must scavenge for food do not always have a healthy diet.

THE DOMESTIC CAT'S WILD RELATIVES

In the wild, members of the cat family are found on every continent of the world except Australia and Antarctica. They live in forests, jungles, grasslands, deserts, or wherever they can find food. However, the increasing development of wild lands by people has endangered many wild cats by reducing their native habitat. Hunting of wild cats for sport and for their beautiful fur has also diminished many species. Without protection, some species of wild cats are in danger of becoming extinct.

Lions are the second largest cat species.

The thick fur of the snow leopard helps to keep it warm.

Most of us will never see tigers, lions, bobcats, or other wild cats in their natural habitats. We can see them in zoos or wild animal parks, however, where we can often observe many of the same behaviors that would occur in the wild. If you go to a zoo and watch a lion yawn, a bobcat bite a piece of meat, or a baby tiger play with its mother's tail, it is not difficult to imagine these animals as oversize versions of your cat at home. In the same way, as you interact with your own cat you can catch glimpses of the wild cat within. Despite their 6,000-year association with humans, cats still retain elements of their wild ancestry and provide us with the opportunity to experience animal behavior up close.

Cheetahs are the fastest of all land animals and can reach speeds of 65 miles per hour in short bursts. Due to hunting, loss of habitat, and a low rate of reproduction, cheetahs are one of the most endangered cat species.

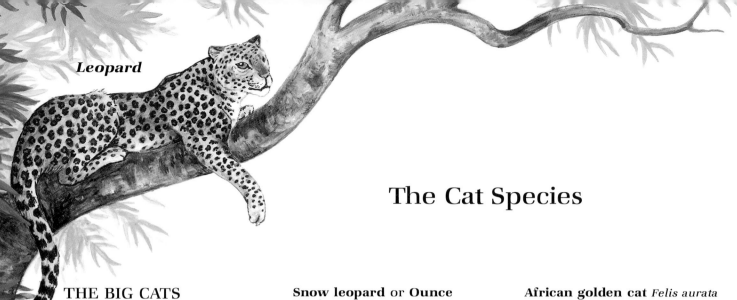

Leopard

The Cat Species

THE BIG CATS

Clouded leopard
Panthera nebulosa

Size: 33-44 pounds
Range: forests of China, Indochina, Sumatra, and Borneo
Endangered; like small cats it can purr and is sometimes classified with the Felis group

Jaguar *Panthera onca*

Size: 126-249 pounds
Range: formerly in southwestern United States, now only in Central and South America
Endangered; the largest cat in the Americas

Leopard *Panthera pardus*

Size: 66-155 pounds
Range: Africa, southern Asia, and China
Endangered; coat has rosette-shaped spots

Lion *Panthera leo*

Size: 270-530 pounds
Range: Africa and India
Endangered in India; the second-largest cat; lives in family groups

Snow leopard or Ounce
Panthera uncia

Size: 80-110 pounds
Range: mountains of central Asia
Endangered; has been successfully bred in zoos

Tiger *Panthera tigris*

Size: 287-573 pounds
Range: Asia and Southeast Asia
Endangered; largest of all the cats; an excellent swimmer and jumper

THE CHEETAH

Cheetah *Acinonyx jubatus*

Size: 86-143 pounds
Range: Africa
Endangered; fastest of all land animals

THE SMALL CATS

Although some scientists have divided the *Felis* genus into several sub-genera, it is treated here as one group.

African desert cat or Sand cat
Felis margarita

Size: 4-5.5 pounds
Range: deserts of Africa and the Near East
Endangered; a close relative of the African wildcat

African golden cat *Felis aurata*

Size: 30-40 pounds
Range: West and central Africa
Forest-dwelling cat

African wildcat *Felis lybica*

Size: 7-13 pounds
Range: Africa
Sometimes classified as a sub-species of the European wildcat

Andean cat or Mountain cat
Felis jacobita

Size: 8-15 pounds
Range: dry mountain regions of South America
Endangered; rarely seen in the wild

Black-footed cat *Felis nigripes*

Size: 2-4 pounds
Range: South Africa
Noted for its fierceness

Bobcat *Felis rufus*

Size: 13-68 pounds
Range: Canada, United States, and northern Mexico
Similar to the northern lynx but smaller

Bornean red cat or Bay cat
Felis badia

Size: 4-7 pounds
Range: Borneo
A relative of the golden cats

Caracal lynx *Felis caracal*
Size: 35-51 pounds
Range: Africa and Asia
More slender than the northern lynx

Chilean forest cat or **Kodkod**
Felis guigna
Size: 4-7 pounds
Range: Chile and Argentina
Related to the ocelot

Chinese desert cat or **Gobi cat** *Felis bieti*
Size: 12 pounds
Range: central Asia
Related to the wildcats

Domestic cat *Felis catus*
Size: 3-18 pounds
Range: worldwide
Varies greatly in size, coat color, and fur length

European wildcat
Felis silvestris
Size: 7-13 pounds
Range: western Eurasia
Similar to the domestic cat

Fishing cat *Felis viverrina*
Size: 12-18 pounds
Range: southern China and Sri Lanka
Related to and larger than the leopard cat

Flat-headed cat *Felis planiceps*
Size: 12-18 pounds
Range: southeast Asia
Endangered; eats fish, frogs, fruits, and sweet potatoes

Geoffroy's cat *Felis geoffroyi*
Size: 4-8 pounds
Range: South America
Related to the ocelot

Jaguarundi *Felis yagouaroundi*
Size: 12-22 pounds
Range: Central and South America, and the southern United States
Endangered; sometimes called the otter cat because of its body shape

Jungle Cat *Felis chaus*
Size: 15-30 pounds
Range: North Africa and Asia
Has tufted ears and a longer head than other wild cats

Leopard cat *Felis bengalensis*
Size: 8 pounds
Range: southern Asia
Endangered; a medium-sized spotted cat

Marbled cat *Felis marmorata*
Size: 12 pounds
Range: eastern Asia, Sumatra, and Borneo
Endangered; coat has distinctive large, rimmed patches

Margay cat *Felis wiedii*
Size: 9-20 pounds
Range: northern Mexico to northern Argentina
Endangered; also called the tree ocelot

Northern lynx *Felis lynx*
Size: 11-64 pounds
Range: northern Eurasia and North America
Has a short "bobbed" tail

Ocelot *Felis pardalis*
Size: 24-35 pounds
Range: North and South America
Endangered

Pallas's cat *Felis manul*
Size: 7-11 pounds
Range: central Asia
Population may be threatened due to overhunting

Pampas cat *Felis colocolo*
Size: 8-14 pounds
Range: southern South America
Related to the ocelot

Puma, cougar, mountain lion, or **panther** *Felis concolor*
Size: 79-227 pounds
Range: North and South America
Endangered; largest of the small cats

Rusty-spotted cat
Felis rubiginosus
Size: 2-4 pounds
Range: Sri Lanka
Related to the leopard cat

Serval *Felis serval*
Size: 30-42 pounds
Range: Africa
Always lives near water

Temminck's golden cat or **Asiatic golden cat**
Felis temmincki
Size: 13-24 pounds
Range: eastern Asia
Endangered; related to the African golden cat

Bobcat

Glossary

domesticate: to shape a species of animal over time to live with and assist humans

dominance hierarchy *(HI-er-ark-ee)*: a social structure that is based on a pecking order in which individual members back down or stand up to one another, depending on their established rank

dominant: being the animal who stands up to another when both are competing for the same resource (such as food or a mate)

feral: living wild after having been domesticated

genus: a biological classification, which ranks between family and species, that groups together animals that have shared characteristics

instinct: an inborn response an animal has to its environment

Jacobson's organ: an organ on the roof of a cat's mouth that combines the senses of taste and smell

papillae *(puh-PIL-ee)*: tiny, rough projections on a cat's tongue

pride: the social group within which lions live

scent glands: specialized cells that produce chemicals that are secreted from the body and create odors

selective breeding: a method of changing or maintaining a kind of animal by the careful selection of parents who have the most desirable traits

sheathed: to be withdrawn into a cover, like a cat's claw being withdrawn into its foot

subordinate: being the animal who backs down to another when both are competing for the same resource (such as food or a mate)

tame: to accustom an individual animal to the presence of humans

tapetum lucidum *(teh-PEET-uhm LU-sehd-um):* a layer of cells in a cat's eye that acts like a mirror, reflecting light into the eye

ultrasounds: sounds above the range of human hearing

vibrissae *(vih-BRIS-ee):* whiskers or hairs that are sensitive to touch or movement

METRIC CONVERSION CHART		
WHEN YOU KNOW:	MULTIPLY BY:	TO FIND:
inches	2.54	centimeters
feet	.3048	meters
miles	1.609	kilometers
acres	.405	hectares
square feet	.09	square meters
ounces	28.35	grams
pounds	.454	kilograms
tons	.9072	metric tons
gallons	3.787	liters

Index

Pages listed in **bold** type refer to photographs.

TOWNSHIP OF UNION
FREE PUBLIC LIBRARY

AAQ-4147